UNDERSTANDING QUANTUM

VOLUME 1 The Universe is Made Up of "Stuff"

Copyright 2015

by

Irwin Tyler

ISBN-13: 978-1515208976
ISBN-10: 1515208974

Ahl Kayn Publishing

Spring Valley, New York
HTTP://www.AhlKaynBooksandArtWork.com

BOOKS BY Irwin Tyler (Yirmi Tyler)

UNDERSTANDING QUANTUM - Volume 2

The Universe Doesn't Make Any Sense

PLANNED FOR PUBLICATION

LATE 2015

UNDERSTANDING QUANTUM - Volume 3

The Theory of Everything

PLANNED FOR PUBLICATION

LATE 2015

POINTS OF HEALTH

The Effectiveness and Safety of

Acupuncture and Acupressure

WHY ACUPUNCTURE? - When Conventional

Medicine Isn't Working As You Hoped

WHY ACUPRESSURE? - When Conventional

Medicine Isn't Working As You Hoped

―――――――――――――――――――――

WHY CHIROPRACTIC? - When Conventional
Medicine Isn't Working As You Hoped

WHY HOMEOPATHY? - When Conventional
Medicine Isn't Working As You Hoped

THE DIET CHOICE PROGRAM - Beat the
Cravings and Enjoy Your Dinner

SO MANY GATES TO THE CITY... A GUIDE
FOR THE MODERN PERPLEXED

**A Book About Jewish Belief and Understanding,
and Making Some Sense Of It**

TARGUM AMERICANA - BERESHIT /
GENESIS

TARGUM AMERICANA – SHEMOT / EXODUS
PLANNED FOR PUBLICATION EARLY
2016

COLLECTING PAPER MONEY WITH
CONFIDENCE

GRADING COINS WITH CONFIDENCE

AVAILABLE AT:

AMAZON.COM

LULU.COM

AHL KAYN PUBLICATIONS WEB SITE

TABLE OF CONTENTS

PROLOG

What is QUANTUM and why does it matter?
To state it very simply:

The discovery of quantum eventually led
to the creation of the atomic and
hydrogen bombs.

Because of the concept of quantum we
have seen the peaceful application of
atomic energy in a satellite continuing on
its journey through outer space that still
sends us its findings although many
years have passed.

Computers and cell phones exist because
of what we understand about quantum.

Our understanding about how this
universe came to be, how it works, and
its ultimate fate rests in great measure on
our understanding of quantum.

Yet, knowing that quantum here refers to a packet of energy that is fixed in amount/size and can not be divided does not seem like such a remarkable idea. On the contrary, its discovery was a monumental revelation. And that is the reason for this book series.

INTRODUCTION

There is no way anyone can explain quantum theory to the beginner or non-scientist by diving right into the subject. It's just too weird if we look only to our experiences in the world around us to try to validate what scientists understand about quantum theory and quantum physics. Instead, we need to sneak up on it, building our knowledge slowly, recognizing anomalies, trying to explain conundrums and seeming impossibilities, testing and discarding and rebuilding ideas, and validating our theories experimentally, bit by bit until at least some of quantum begins to make sense.

This book seeks to make enough sense to satisfy you without the need for hard physics and mathematics.

You will likely have as many questions at the end of this book series as you have now at its beginning. Well, that's OK. Physicists and mathematicians studying quantum theory in depth have faced the same situation throughout their careers.

Although you will still have questions, the questions at the end will be different from those you had at the beginning because you will have learned a lot. Some will be questions that a future, deeper study of quantum may be able to answer for you. And some may be questions that no one in our lifetime will confidently be able to answer.

Understanding quantum is still a journey in progress.

Irwin Tyler   2015

VOLUME 1 The Universe is Made Up of "Stuff"

Our earliest recording of man's inquiry into the nature of "things" comes from Greek philosophical writings of around 1000 BCE. Certainly, there were other societies and earlier societies that also thought about the nature of "things" but it is through the Greek lineage that scientific inquiry and the current understanding of quantum theory springs.

Greek thought was based upon what people could experience, which became the basis for reasoning about the implications of these experiences. They observed permanence and change, stability and movement, unique objects and many objects as a group called by a single name (Fido the dog or all dogs), and they asked: what makes the world what it is? What makes things appear the same yet are different? Things move, but why is the movement of a rock thrown in the air different from the movement of a leaf falling from a tree, or is it? What makes the substance of rock different from the substance of skin? And what is an idea?

In the 5th Century BCE Leucippus of Miletus and his pupil Democritus asked the following question:

What happens if you break a piece of matter in half and in half again. Then do it again and again. Is there a limit to how many breaks you can complete before it can break no further?

They reasoned that at some point there must be an end to the process, where it leaves an unbreakable bit of matter, which Democritus called an atom. Once they reached this conclusion they attempted to explain what these atoms were. They reasoned the following:

1. All matter consists of invisible, undividable particles called atoms.

2. Atoms are indestructible.

3. Atoms are solid but invisible.

4. Atoms are homogenous – an atom is a single substance.

5. There is empty space surrounding all atoms.

6. Atoms differ in size, shape, mass, position, and arrangement.

- *Solids are made of small, pointy atoms.*
- *Liquids are made of large, round atoms.*
- *Oils are made of very fine, small atoms that can easily slip past each other.*

About 100 years later Aristotle, with a different reasoning about the nature of substance, concluded, quite differently, that all substances were made up of varying combinations of air, fire, water and earth. So respected were his opinions and teachings that the ancient world rejected the ideas of Democritus and held to his alternative view of matter.

Because Democritus, in developing his philosophy of reality, proposed the idea that the universe had always been in existence and that there would be no end to it, the Catholic Church debated and rejected atomism and retained as "truth" Aristotle's concept, calling Democritus' theory of the atom an idea that was equivalent to proposing Godlessness. This position held sway until the 1800's, resulting in no significant serious scientific inquiry being pursued throughout this period of 2100 years.

Matter is Solid

In the Age of the Enlightenment, that 100+ year period leading to the 1800's, cultural and intellectual forces in Western Europe began to emphasize reason, analysis, and individualism rather than the authority of church or monarch. This led to a growing number of philosophers and experimenters reviving inquiry into the nature of the universe and its parts.

At the beginning of the 19th Century an English chemist, John Dalton, performed numerous chemical experiments. He was able to demonstrate that matter, as Democritus had proposed, seemed to consist of elementary lumpy particles (atoms). Through his experiments Dalton further deduced and expanded his atomic theory:

- Elements consist of tiny particles called atoms.
- Atoms are solid and indivisible.

- Atoms can be neither created nor destroyed.

- An element is one of a kind (pure) because all atoms of an element are identical.

- All the atoms that make up an element have the same mass (the amount of matter each atom consists of).

- All elements are different from each other because their atoms have different masses.

- A compound is a combination of elements bonded together.

- Chemical reactions involve the rearrangement of combinations of atoms.

Experiments proliferated throughout the 19th Century. English physicist J. J. Thomson made a critical discovery of the electron, a negatively charged particle, in 1897.

It had been observed that certain materials, when rubbed, developed a difference in charge between the two materials. Particles from one of these materials could be "pulled off" and sent through a tube emptied of air. He measured the mass of these particles, which he determined had negative charges, and found that they had a mass very much smaller than the mass of a hydrogen atom.

Thomson theorized that atoms were made up of collections of these negative particles, now known as electrons. But he also observed that atoms had no charge. He then reasoned that atoms had another group of particles with positive charges to balance out the electrons.

Since it was still thought that atoms were solid and indivisible, in 1900 Thomson proposed the Plum Pudding model of the atom. He concluded that atoms are made of positive cores, the pudding, and negatively charged particles, like bits of plum or raisins, embedded within the pudding.

So, by the turn of the 20th century matter was thought to be made up of atoms, solid balls yet formed from two oppositely charged kinds of materials. An atom of each element had a different mass from that of another element, meaning that each element had a different number of these negatively charged particles and positively charged other "stuff". Elements were made up of groups of the "Plum Pudding" atoms stuck together.

Questions still remained. Were the embedded negative particles stationary or moving? If they moved, what paths did they take? Were such paths fixed, random, or changeable?

At the same time, others were performing seemingly unrelated experiments, working with materials and equipment that others were not directly interested in, and making new discoveries. Many resulted in new unanswered questions. Few could realize that some of these other results and ideas would begin to answer their earlier questions.

Matter is Not So Solid

There is no end of amazement at the human brain's ability to propose questions, to question answers, and to devise ways to investigate the world around them.

Henri Becquerel, Pierre and Marie Curie, and Ernest Rutherford each were fascinated by the discovery that certain elements (uranium, thorium, radium, and polonium) spontaneously radiated some kind of energy. It was at first thought to be a form of x-ray. Rutherford's experiments showed that there actually were at least two kinds of radiation. Two that he clearly identified he called alpha rays and beta rays.

Based on earlier experiments with uranium by Becquerel, who used various barriers between uranium and photographic plates to study characteristics of the emitted radiation, Rutherford showed that the alpha rays were less powerful than the beta rays. Somehow, beta rays were able to punch their way through a greater thickness of aluminum than the alpha rays were able to achieve.

Cathode rays had been discovered in 1876. It was known that these rays could be manipulated by electrical devices. Rutherford was able to measure the mass of cathode rays, thus showing that these rays were actually particles. When Rutherford, in his 1897 experiments with beta rays, applied a magnetic field to his experimental apparatus he observed that the beta rays shifted their path. This meant that beta rays were not like x-rays, which would travel a straight path, but were charged particles just like cathode rays. It was soon determined that the beta rays were, in fact, a stream of the recently discovered electron.

Meanwhile, the Curies, experimenting with alpha rays, found that they behaved like a projectile, losing energy as they travelled through thicker and thicker sheets of aluminum. When Rutherford finally used powerful enough magnets in his experiments he was able to observe that alpha rays, like beta rays, could be made to shift their paths. It was known that actual rays could not be affected by magnets. Evidence was now conclusive that alpha rays actually were particles, and much heavier than electrons. This was why he needed more powerful magnets to deflect them. He was now also able to show that these particles had an opposite charge to that of the electron.

Could the Plum Pudding theory of the structure of the atom, a ball of positive particles studded with electrons, explain these new discoveries?

Things were moving fast in the world of experimental physics in the late 19th and early 20th Century years. As we will see, the Plum Pudding theory, as convincing as it was, did not last even a decade.

The Universe is Made Up of Hardly Anything

In the absence of a magnetic field Rutherford bombarded a sheet of gold foil with alpha particles. Most particles passed straight through. Unexpectedly, he observed that an occasional alpha particle changed direction sharply from its original path, sometimes bouncing straight back from the foil. His explanation for this experimental result is as follows:

- Alpha particles are solid.
- Alpha particles can go through gold foil.
- There must be spaces between the atoms of gold foil to allow the alpha particles to pass through.
- Since the atom's "pudding" was reasoned to be positively charged, and the alpha particle was experimentally found to be positively charged, the alpha

particles must bounce back because the two positive charges repel each other.

- Since the atom also contains electrons (the raisins or plum bits), if the electron was embedded within the central "pudding" the whole would be electrically neutral. But the experiment shows this not to be the case.

- The only place for the electron to occupy is some small part of the empty space between atoms, and that space must be large enough to allow passage of an alpha particle.

- This means that there is a positive core concentrated into a small region with a large amount of empty space around it, and then some arrangement of electrons further out.

In the 1890s Max Planck had performed a series of heat experiments. As he heated metals they glowed in different colors, and each color radiation had a different amount of energy. In 1900 he showed that atoms absorbed this energy in specific amounts. To illustrate, an atom might absorb 2 units of energy or 3 units of energy but not 2.5 units. He called these units "quanta".

Planck found that the concept of "quanta" could not be explained by any of the laws of physics then known. Despite the experimental demonstrations and mathematical "proofs" that quanta existed, for many years Planck, himself, found the concept of quanta intellectually illogical, so strong was the pull of classical physics as it had evolved through the 19th Century. That there was empty space in the atom, and that atoms could absorb energy only in specific amounts were just two of the many new revelations about the nature of matter that still had no explanation.

Matter has Structure

The most important outcome of Rutherford's experiments was that they revealed that the atom had a nucleus. By 1909 he had concluded that the bulk of the mass of an atom was in its nucleus. Based on his experiments with alpha particles, Rutherford speculated that the negative electrons orbited a positive center much like the solar system, where the planets orbit the sun.

How much space there might be and the kinds of orbits electrons might travel was still a matter of speculation. He proposed that the orbits of electrons were varied and scattered, creating a kind of cloud of electrons about the nucleus.

By 1920 Rutherford was able to further refine his understanding of the atom's nucleus. He was convinced that there were both positive particles (protons) and a mysterious neutral particle together making up the nucleus.

Since his experiments in shattering the nucleus produced electrons, his thought was that each electron was somehow tightly bound to a proton, creating a neutral particle, a neutron. Thus, his model of the nucleus consisted of a number of protons bound to an equal number of electrons which balanced and cancelled out their charges, plus additional protons that would account for the measured positive charge of the nucleus.

His theory seemed to cover many open questions: he could account for the heavy weight of the nucleus, its positive charge, and his experimental results.

Matter is Almost Empty Space

Earlier, in 1913, Rutherford bombarded nitrogen and several other light elements with alpha particles. He observed that some nuclei could be shattered, with the collision producing fast-moving protons (positively charged particles). But most alpha particles passed through. So, he concluded, the nucleus was mostly empty space.

At this point atoms were known to be neutrally charged, with a proposed nucleus and some kind of arrangement of orbiting electrons and lots of empty space.

But the nucleus needed to be positively charged to balance the negative electrons. The nucleus must, therefore, consist of positively charged particles. These positively charged particles had to be much heavier than the electrons, since they repelled the "large" alpha particles.

The "Plum Pudding" concept of the atom was officially dead. But the picture of the atom still was not complete.

What about all this empty space? Where were the orbiting electrons and why didn't they just fly off into space?

Rutherford's student, Neils Bohr, in 1919 examined earlier experiments and noted logical inconsistencies in proposed theories about the atom. He accepted the generally believed idea that the atom was solar system-like, with electron "planets" orbiting a nucleus "sun", but he uniquely also proposed that the electrons did not merely orbit in a somewhat random "cloud" anywhere outside the nucleus but actually orbited in some fixed arrangement around the nucleus.

In the 1850's scientists, in heating pure elements until they glowed, radiated light (energy) only at specific frequencies/colors. Every element had its specific pattern of frequencies of emitted light. Why this was so could not be explained by the then known principles of physics and theories of the nature of the atom.

Bohr reasoned about these unexplained results and of other unexplainable results in previous experiments with light. He concluded that an orbiting electron absorbed heat energy by shifting its orbit to a higher energy level. When each electron returned to its originally lower energy level it radiated this energy back in the form of light at a specific frequency, representing the difference in the energy contained at each level.

But the supposed randomness of the orbiting electrons did not support all of the experimental results. The specific frequencies could occur only if there was a fixed arrangement of electron orbits. Physicists now had still another new understanding about the nature of atoms.

What Holds Matter Together?

Experiments of all kinds continued, and in 1930, when Bothe and Becker bombarded beryllium with alpha particles, this produced both the expected charged particles but also an unexpected neutral, uncharged radiation.

Calculations of the energy of all the radiation plus particles released in these collisions showed that the neutral radiation could not be gamma rays, as had been thought at first. Further bombardment experiments by Chadwick in 1932, using different materials, resulted in his determining that this neutral radiation was, in fact, another particle, with a mass close to that of the proton.

By this time it was clear that tiny differences between theoretically calculated energy levels, and particle masses that differed significantly from those measured in carefully constructed experiments, actually were important. These differences hinted that the theories producing the calculations either were incomplete or actually partially or fundamentally wrong. This new understanding would assume greater importance in the future as scientists sought to verify and explain the results of older experiments.

While it was now seen that the atom's nucleus consisted of protons and neutrons packed together, nothing in the experiments changed the understanding of twenty years earlier that the nucleus consisted mostly of empty space.

Nothing as yet could explain why and how the presumed solar system-like structure of the atom could exist and sustain itself. If negative electrons moved around the positive nucleus, why did electrons not crash into the nucleus? Electrons were known to somehow exist within the nucleus: did these electrons also move? Why do they remain in the nucleus? Why do protons in the nucleus not repel each other instantaneously?

Each new finding seemed to lead to an endless series of new questions and new experiments. And each new finding disclosed ever tinier "particles" and strange new "things". Three thousand years of serious inquiry still had left the nature of "nature" largely a mystery.

END OF VOLUME 1

Your online comments and your comments directly to the author will be appreciated.

BIBLIOGRAPHY

http://www.nobeliefs.com/atom.htm

http://the-history-of-the-atom.wikispaces.com/Democritus

Philosophy and Philosophers: An
   Introduction to Western Philosophy
      By John Shand, Published 2014 by
         Routledge, London and New
         York

http://www.chemteam.info/Radioactivity/Disc-of-Alpha&Beta.html

http://www.chemteam.info/Radioactivity/Disc-Alpha&Beta-Particles.html

https://en.wikipedia.org/wiki/Atomic_orbital

http://hyperphysics.phy-
astr.gsu.edu/hbase/Particles/neutrondis.html

The Teaching Company: Science in the Twentieth Century

Lecturer: Steven L. Goldman

AUTHOR BIOGRAPHY

Near the beginning in his career with IBM Irwin Tyler revealed a diverse talent for engineering and the sciences (an earlier degree in Civil Engineering and subsequent NY State License in Civil Engineering), for software system design and development, and for advanced research (issued a U.S. Patent in Artificial Intelligence). When IBM discovered his writing ability they provided him a platform for writing several marketing guides for the company, as well as a multimedia presentation.

It was during this period that he wrote, produced, and directed two local plays with theological themes.

After Irwin's retirement from IBM he truly found his writing calling, producing to this point ten books (visit his web site http://AhlKaynBooksandArtWork.com), more than forty published Articles and Letters to the Editor (including Science News, Newsmax, In These Times, USAA Magazine) and has developed a growing career as a ghost writer.

His ability to make the complex understandable without compromising accuracy is a hallmark of his works.

INDEX

www.ingramcontent.com/pod-product-compliance
Lightning Source LLC
Chambersburg PA
CBHW071014180526
45168CB00003B/1423